ELECTRICITY CITY

A Problem-Based Unit

The College of William and Mary
School of Education
Center for Gifted Education
Williamsburg, Virginia 23185

18877

ELECTRICITY CITY

A Problem-Based Unit

The College of William and Mary
School of Education
Center for Gifted Education
Williamsburg, Virginia 23185

Center for Gifted Education Staff:
Project Director: Dr. Joyce VanTassel-Baska
Project Managers: Dr. Shelagh A. Gallagher
Dr. Victoria B. Damiani
Project Consultants: Dr. Beverly T. Sher
Linda Neal Boyce
Dana T. Johnson
Dr. Jill T. Burruss
Donna L. Poland
Dennis R. Hall

Teacher Developer:
Judi Ellis

funded by Jacob K. Javits Program,
United States Department of Education

KENDALL/HUNT PUBLISHING COMPANY
4050 Westmark Drive   Dubuque, Iowa 52002

SCI
C686e
1997

CONTENTS

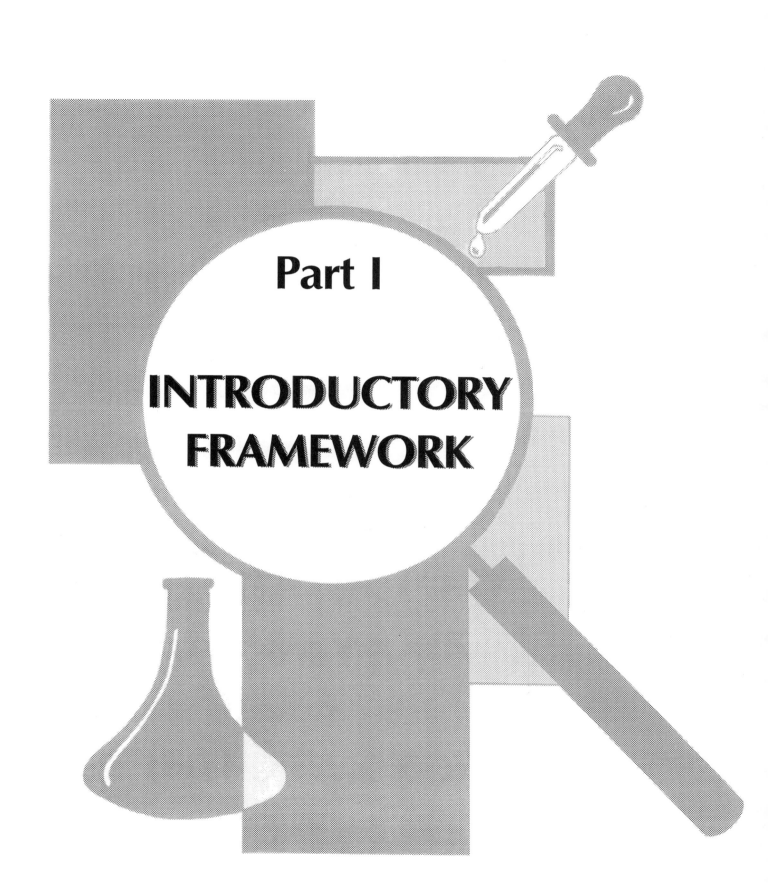

Part I

INTRODUCTORY FRAMEWORK

INTRODUCTION

Electricity City is a problem-based science unit designed for high ability learners. It has been used successfully with all learners in a wide variety of situations, from pull-out programs for gifted learners to traditional heterogeneously grouped classrooms. It allows elementary students to explore models, electricity, and circuits in a novel way, namely through the process of grappling with an ill-structured, "real-world" problem.

Because the unit is problem-based, the way in which a teacher implements the unit will necessarily differ from the way in which most traditional science units are used. Preparing for and implementing problem-based learning takes time, flexibility, and a willingness to experiment with a new way of teaching.

The total time required for completion of *Electricity City* should be minimally 30 hours, with more time required for additional activities.

RATIONALE AND PURPOSE

This unit has been designed to introduce fifth and sixth grade students to electricity in an engaging fashion. The problem-based learning format was chosen in order to allow students to acquire significant science content knowledge in the course of solving an interdisciplinary, "real-world" problem. This format requires students to analyze the problem situation, to determine what information they need in order to come up with solutions, and then to find that information in a variety of ways. In addition to library work and other information-gathering methods, students, with teacher facilitation, will perform experiments of their own design in order to find information necessary to come up with and evaluate solutions to the problem. The problem-based method also allows students to model the scientific process, from the problem-finding and information-gathering steps to the evaluation of experimental data and the recasting or solution of the problem. Finally, the overarching scientific concept of systems provides students with a framework for the analysis of both their experiments and the problem as a whole.

GOALS AND OUTCOMES

➡ To understand the concept of systems

Students will be able to analyze several systems during the course of the unit. These include the "problem system," defined by the boundaries of an electrical circuit and real-world systems, such as the electrical system of a city or complex building. In addition, all experiments set up during the course will be treated as systems.

Systems Outcomes

A. For each system, students will be able to use appropriate systems language to identify boundaries, important elements, input, and output.

B. Students will be able to analyze the interactions of various system components with each other and with input into the system, both for the real-world systems and for the experimental systems.

C. Based on their understanding of each system's functioning, students will be able to predict the impact of various approaches to solving the electrical problem.

D. Students will be able to transfer their knowledge about systems in general to a newly encountered system. In the final assessment activity, students will be given a new system to analyze in the same way that they have analyzed the systems in the unit.

➡ To design scientific experiments necessary to solve given problems

In order to solve these scientific problems, students will be able to design, perform, and report on the results of a number of experiments.

Scientific Process Outcomes

A. Students will be able to explore a new scientific area, namely electricity.

B. Students will be able to identify meaningful scientific problems for investigation during the course of working through the electricity problem and its ramifications. The problem includes the behavior of electrical circuits and the nature of electricity and its generation.

C. In order to answer these scientific questions, students will create and test electrical circuits of their own design.

D. During their experimental work, students will:

—Demonstrate good data-handling skills

—Analyze any experimental data as appropriate

—Evaluate their results in light of the original problem

—Use their enhanced understanding of the area under study to make predictions about similar problems whose answers are not yet known to the student

—Communicate their enhanced understanding of the scientific area to others

Students will be able to use electrical components to create simple circuits that work and that fill a particular need; they will also be able to predict the behavior of electrical circuits designed by others.

SPECIFIC CONTENT OUTCOMES

A. Students will be able to correctly use terms related to electricity and circuits (AC/DC, resistor, conductor, insulator and so on).

B. Students will be able to describe the properties of different circuit elements, including power sources, wires, resistors, and switches.

C. Students will be able to describe the flow of electricity in a circuit.

D. Students will be able to read and create simple circuit diagrams.

E. The student will be able to use circuit elements to build complete electrical circuits with specified properties.

F. Students will be able to determine whether a circuit is complete or incomplete, either by looking at a circuit diagram or by devising a way to test the circuit directly.

G. Students will be able to describe the creation and movement of electricity in an electrical power system from the source to the consumer.

H. Students will apply their understanding of electricity and electrical systems to the creation of a model electrical system.

I. Students will be able to describe and apply the concept of scale during the planning and construction of their model.

J. Students will understand hazards associated with electricity and the general procedures for dealing with electrical emergencies.

ASSESSMENT

This unit contains many assessment opportunities that can be used to monitor student progress and assess student learning. Opportunities for formative assessment include:

• The student's problem log, a written compilation of the student's thoughts about the problem. Each lesson contains suggested questions for students to answer in their problem logs. The problem log should also be used by the student to record data and new information that they have obtained during the course of the unit.

- Other report forms are used to help the student explain their solutions to particular parts of the problem.
- Teacher observation of student participation in large-group and small-group activities.

Opportunities for cumulative assessment include:

- The final resolution activity, which involves a small group presentation of a solution to the unit's ill-structured problem; the quality of the solution will reflect the group's understanding of the science involved as well as the societal and ethical considerations needed to form an acceptable solution.
- Final unit assessments, which allow the teacher to determine whether individual students have met the science process, science content, and systems objectives listed in the Goals and Objectives section at the beginning of the unit.

SAFETY PRECAUTIONS TO BE TAKEN IN THE LAB

As this unit involves laboratory work, some general safety procedures should be observed at all times. Some districts will have prescribed laboratory safety rules; for those that do not, some basic rules to follow for this unit and any other curriculum involving scientific experimentation are:

1. Students must behave appropriately in the lab. No running or horseplay should be allowed; materials should be used for only the intended purposes.
2. No eating, drinking, or smoking in lab; no tasting of laboratory materials. No pipetting by mouth.
3. If students are using heat sources, such as alcohol burners, long hair must be tied back and loose clothing should be covered by a lab coat.
4. Fire extinguishers should be available; students should know where they are and how to use them. Attention should be given to the difference between fire extinguishers for electrical fires and those for other types of fires.

Some specific safety rules relevant to implementing this unit: This unit does use circuit components, which can be dangerous if used inappropriately. Power sources supplied to students should consist of small batteries (such as D-cell batteries) rather than large power supplies; the use of capacitors should be avoided, as charged capacitors can give nasty shocks. The teacher should also be prepared to demonstrate the proper use of all circuit components and to oversee the hands-on technical work that students perform. Wire cutters and wire strippers are tools that are used in this unit but must be used with great care; teachers may want to cut and strip the wire themselves.

X-acto knives and hot glue guns are useful for building models but can be dangerous for students. Teachers may want to do the cutting and gluing for students. Students should not play with or put anything into a household circuit.

MATERIALS LIST

Materials needed for each individual lesson are listed in the "Materials and Handouts" section of the lesson.

LESSON FLOW CHART

Problem-based learning is not easy to plan, because it is driven by student questioning and interest. We have included estimated durations for each lesson in this unit, but be prepared to be flexible and to move with the students. We have also included a diagram (Figure 1) which shows the relationship between the individual lessons and experiments suggested in the unit. In general, lessons shown higher in the diagram are prerequisites for those shown lower in the diagram. Be aware that this diagram may not reflect all of the time that you will need to spend; students may well come up with unanticipated, yet valid, experiments or lines of questioning.

We feel that some of the lessons are essential for all students, while others can be done with a subgroup as long as the subgroup reports its results back to the whole class.

TAILORING *ELECTRICITY CITY* TO YOUR LOCATION

Classroom experience during the unit piloting process has shown that this unit is *much* more powerful when tailored for the location in which it is being presented. Accordingly, consider the following suggestions:

1. This unit needs to be personalized to your town, city, or locality. The design of the recreational complex in the problem statement can be altered to suit your locale.

2. When addressing the storm-related problem given in Lesson 1, the use of local maps and information is appropriate in the subsequent lessons.

3. Involve local experts (electricians, electrical line foreman, electrical company workers, architects, etc.) as speakers and on-going resources in the problem-solving process.

4. Work with librarians to plan the unit and to assist students in finding information. In addition to school librarians and academic librarians, special libraries (museums, corporations, historical societies, etc.) offer vast resources relevant to the unit.

FIGURE 1

LESSON FLOW CHART

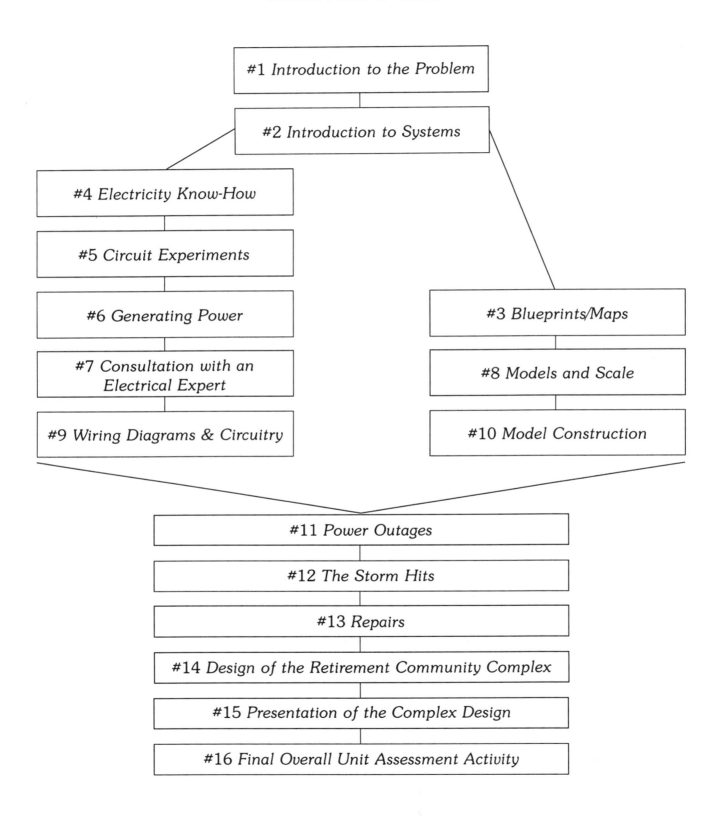

#1 Introduction to the Problem

#2 Introduction to Systems

#4 Electricity Know-How

#5 Circuit Experiments

#6 Generating Power

#7 Consultation with an Electrical Expert

#9 Wiring Diagrams & Circuitry

#3 Blueprints/Maps

#8 Models and Scale

#10 Model Construction

#11 Power Outages

#12 The Storm Hits

#13 Repairs

#14 Design of the Retirement Community Complex

#15 Presentation of the Complex Design

#16 Final Overall Unit Assessment Activity

GLOSSARY OF TERMS

Alternating Current (AC): Current in which the electrons reverse their direction regularly.

Black-out: Lack of illumination caused by an electrical power failure.

Brown-out: A reduction or cutback in electric power, especially as a result of a shortage, a mechanical failure, or overuse by consumers.

Boundary of a System: Something that indicates or fixes a limit on the extent of the system.

Charge: Physical property of matter that can give rise to an electric force of attraction or repulsion.

Circuit: Complete path through which electricity can flow.

Conductor: Material which permits electrons to flow freely.

Current: Flow of charge; measured in units called amperes.

Direct Current (DC): Current consisting of electrons that flow constantly in one direction.

Electric Motor: Device that uses an electromagnet to convert electrical energy to mechanical energy that is used to do work.

Electrical Generator: A device that converts mechanical energy into electrical energy by moving electrical conductors in the presence of a magnet.

Element of a System: A distinct part of the system; a component of a complex system (a subsystem).

Generator: Device that uses electromagnets to convert mechanical energy to electrical energy.

Input from a System: Something that is put in the system; an addition to the components of the system.

Insulator: Material made up of atoms with tightly bound electrons that are not able to flow freely.

Model: A structural design or miniature representation.

Ohm's Law: Expression that relates current (I), voltage (V), and resistance (R): $V = I \times R$.

Output from a System: Something that is produced by the system; a product of the system interactions.

Parallel Circuit: Circuit in which different parts are on separate branches; if one part does not operate properly, current can still flow through the others.

Potential Difference: Difference in charge as created by opposite posts of a battery.

Power Surge: A sudden increase in the amount of current flowing through a power line.

Resistance: Opposition to the flow of electric charge; measured in units called ohms.

Resistor: A device used to control current in an electric circuit by providing resistance.

Scale: The size of a sample in proportion to the size of the actual items, as on maps or models.

Scientific Process (or Research): The scientific research process can be described by the following steps:

1. Learn a great deal about your field.
2. Think of a good (interesting, important, and tractable) problem.
3. Decide which experiments/observations/calculations would contribute to a solution of the problem.
4. Perform the experiments/observations/calculations.
5. Decide whether the results really do contribute to a better understanding of the problem. If they do not, return to either step 2 (if you're very discouraged) or step 3. If they do, go to step 6.
6. Communicate your results to as many people as possible. If they're patentable, tell your lawyer before you tell anyone else, and write a patent application or two. Publish them in a scientific journal (or if they are really neat, in *The New York Times*); go to conferences and talk about them; tell all of your friends.

Series Circuit: Circuit in which all parts are connected one after another; if one part fails to operate properly, the current cannot flow.

Short Circuit: A circuit in which the electrons take a shorter path back to the power source rather than follow the usual longer path through a light bulb or other appliance.

System: A group of interacting, interrelated, or interdependent elements forming a complex whole.

Transformer: Device that increases or decreases the voltage of alternating current.

Transistor: Device consisting of three layers of semiconductors used to amplify an electric signal.

Turbine: Engine turned by the force of gas or water on fan blades.

Voltage: Potential difference; energy carried by charges that make up a current; measured in units called volts.

Watt: SI unit of power defined as one joule per second (W = J/S). Named after James Watt who greatly improved the rate at which steam engines could do work. In electricity one watt is defined as one volt times one ampere (W = V x A). In other words, a watt is a measure of the current times electrical potential.

LETTER TO PARENTS

Dear Parent or Guardian:

Your child is about to begin a science unit that uses an instructional strategy called problem-based learning. In this unit students will take a very active role in identifying and resolving a "real-world" problem constructed to promote science learning. Your child will not be working out of a textbook during this unit but will be gathering information from a variety of other sources both in and out of school.

The goals for the unit are:

- *To understand the concept of "systems."*
  Students will be able to analyze several systems during the course of the unit. These include the "problem system," defined by the boundaries of an electrical circuit and real-world systems, such as the electrical system of a city or complex building. In addition, all experiments set up during the course will be treated as systems.

- *To apply the principles of electrical circuitry.*
  Students will be able to use electrical components to create simple circuits that work and that fill a particular need; they will also be able to predict the behavior of electrical circuits designed by others.

- *To design scientific experiments necessary to solve given problems.*
  In order to solve scientific problems, students need to be able to design, perform, and report on the results of a number of experiments. During their experimental work, students will:

  —Demonstrate good data-handling skills

  —Analyze any experimental data as appropriate

  —Evaluate their results in the light of the original problem

  —Use their enhanced understanding of the area under study to make predictions about similar problems whose answers are not yet known to the student

  —Communicate their enhanced understanding of the scientific area to others

Since we know from educational research that parental involvement is a strong factor in promoting positive attitudes toward science, we encourage you to extend your child's school learning through activities in the home.

Ways that you may wish to help your child during the learning of this unit include:

- Discuss systems, including family systems, educational systems, etc.
- Discuss the problem they have been given.

- Engage your child in scientific-experimentation exercises based on everyday events such as: In a grocery store, how would you test whether it's better to go in a long line with people having few items or a short line with people having full carts?

- Take your child to area science museums and the library to explore how scientists solve problems.

- Use the problem-based learning model to question students about a question they have about the real world, e.g., How does hail form? Answer: What do you know about hail? What do you need to know to answer the question? How do you find out?

Thank you in advance for your interest in your child's curriculum. Please do not hesitate to contact me for further information as the unit progresses.

Sincerely,

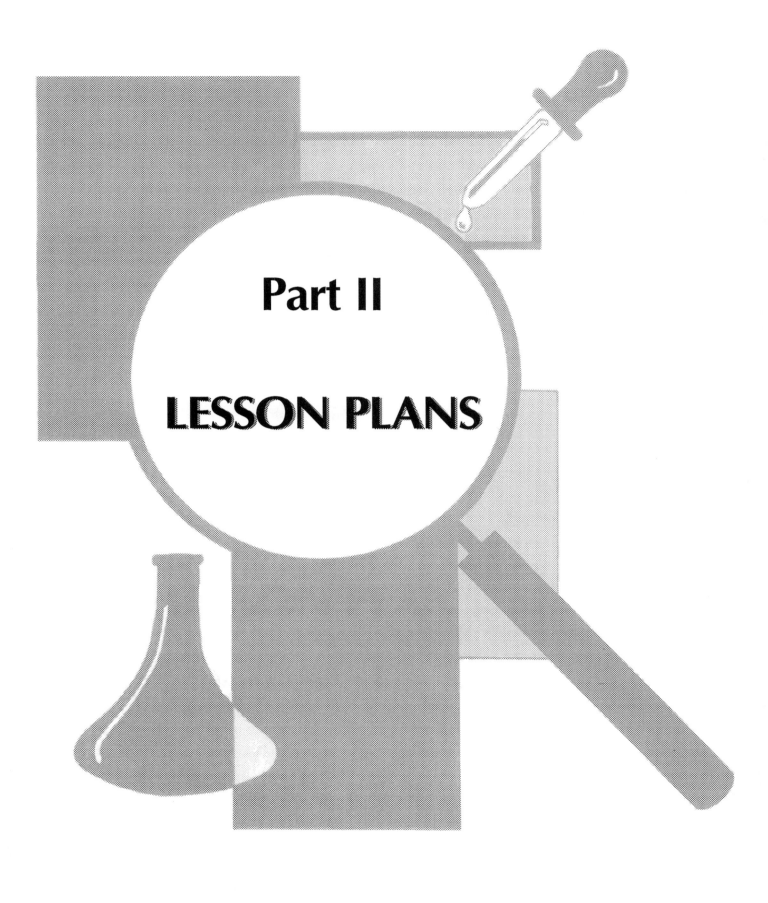

Part II

LESSON PLANS

Introduction to the Problem

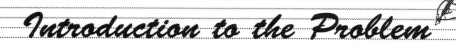

LESSON LENGTH: Two or three sessions

INSTRUCTIONAL PURPOSE

- To introduce students to the central problem of the unit.

MATERIALS AND HANDOUTS

Pencil and paper (optional)

City pamphlets and maps

Handout 1.1: Problem Statement

Handout 1.2: "Need to Know" board

Handout 1.3: Additional Problem Statement Information

Handout 1.4: Group Activity Chart

Handout 1.5: Problem Log Questions

Session 1

THINGS TO DO

1. Introduce the problem to students and have them identify what they know about the problem. Distribute Handouts 1.1 and 1.2.

2. Next, ask students to list what they would need to know in order to solve the problem. Use a "Need to Know" board to chart what students know about the problem, what they need to know, and how they can find out that information. As students list what they need to know, also have them state what they think the problem is. This board can be used throughout the unit, revising information as it changes during the course of the unit.

3. Introduce the Additional Problem Statement Information (Handout 1.3).

4. Ask students to speculate about the things a "linesman" (or repair person) has to consider as he works. Inform students that they will be working on their problem in the classroom; ask them how they would get the job done in that setting. (**Note:** If they don't suggest it, prompt them to talk about the possibility of using a model.)

5. Divide students into collaborative work groups (4 or 5 to a group). Students should work in these groups to brainstorm places of interest in the city for which they might want to build a model.

THINGS TO ASK

- Now what's going on? What do you know about the situation?

- What are the things that "linesmen" or repair persons know about?

- What kinds of problems cause power outages?

- Which of these could be the source of the expected problems? Why do you think so?

- How can you find out the necessary information in order to get started on your temporary assignment?

- What information do you need to know first and why do you need to know it?

- How could we actually replicate this in our classroom?

Session 2

THINGS TO DO

1. Divide students into collaborative learning groups (as in Session 1). This group will work together throughout the duration of the unit. Have the small groups share their lists with one another and select a particular area of the city to be responsible for repairing. Ask students to list the materials they will need to begin work constructing their map.

2. City maps and tourist pamphlets should be available for the groups as reference material. By the end of this session each group should decide which area of the city they would like to investigate, model, and trouble shoot.

THINGS TO ASK

- In order to complete this task what would you need to know?

- What do you already know that could help you?

- How could you find out what to use to model an electrical system?

- How can you find out about the physical structure of your area? About its demand for electricity?

- What else (other than electricity) is important for this area?

ASSESSMENT

1. Problem Log Questions on problem definition.
2. Quality of thought indicated on Group Activity Chart (Handout 1.4).

NOTE TO TEACHER

The initial problem statement sets the stage for students to start asking questions about electricity and various systems. However, the Additional Problem Statement Information (Handout 1.3) is what actually drives the students' investigation into various fields of study. Most of the lessons in the unit address the additional problem—an urgent situation in which the students must educate themselves to the city, its "systems," and electrical power. They will build a model of a section of the city and wire it. After learning about these subjects and related aspects, students address the original problem, to design a recreational complex, during the final lessons. The complex design should serve as an assessment activity that integrates everything students learned in the prior lessons.

You may decide to have students make a model of a single building such as your school instead of a section of the city.

HANDOUT 1.1

PROBLEM STATEMENT

You are a newly hired employee for the power company. Your first task after completing orientation training is to work on a team developing an exclusive recreational complex to be built in the center of town. It is a big project but it has federal and state funding. Your role is to make certain that the power (electricity) requirements are planned appropriately and met. You are also charged with seeing that backup plans are available and adequate for this facility which will serve the needs of all community groups including senior citizens and special needs individuals. Your college training stressed city management and planning, not electricity.

HANDOUT 1.2

"NEED TO KNOW" BOARD

What do we know?	What do we need to know?	How can we find out?

HANDOUT 1.3

ADDITIONAL PROBLEM STATEMENT INFORMATION

It is day three of the orientation training session. Yesterday's session was disrupted by the news of an 8-state power outage that left more than a million people without power in another region of the country. Now there is news of a hurricane with accompanying tornadoes bearing down on your region. The instructor warns you that if there are major outages, you and the other new employees will be called on to help get the system back on line and restore electricity to homes and businesses. That means that you will act as an assistant to grid workers, "linesmen," electricians, and general repair people. You have no background in the area of repairing damage or restoring power and you don't know what to do. The rest of the group is as clueless as you are. The weather forecast doesn't look promising. How can you be prepared? The instructor is willing to change the sessions if you want. What should you do?

HANDOUT 1.4
GROUP ACTIVITY CHART

1. List and discuss some interesting places in the city. What electrical problems might occur here if a storm hits? Choose one that your group would like to investigate and repair.

2. What do you need to know about electricity if you are to investigate one of these problem areas?

3. What are some things an individual working with electricity needs to consider about electrical problems?

HANDOUT 1.5

PROBLEM LOG QUESTIONS

1. After our discussion, what do you think the problem is?

2. Why do you think this is the main problem?

3. Is it the same problem you thought it was when we first started talking?

4. How has your perception changed since you first saw the problem?

5. What area seems most interesting to you right now? Why?

Introduction to Systems

LESSON LENGTH: One session

INSTRUCTIONAL PURPOSE

- To introduce the systems concept.

MATERIALS AND HANDOUTS

Chart paper

Handout 2.1: System Parts Chart

THINGS TO DO

1. Discuss the concept of "systems" with students by introducing the terms associated with systems and have students define each term:

boundary:	something that indicates or fixes a limit on the extent of the system
elements:	a distinct part of the system; a component of a complex entity
input:	something that is put in the system; an addition to the components of the system
output:	something that is produced by the system; a product of the system interactions
interactions:	the nature of connections made between/among elements and inputs of a system

2. Elicit examples of systems with which students are familiar. Have them identify and discuss the parts of those systems.

3. Discuss whether or not a city could be considered a system.

4. List the components that make the city a system. Have students complete System Parts Chart (Handout 2.1) for the city system.

THINGS TO ASK

- Is there a boundary to the city? What is it?
- What are some of the elements in the city? Can these elements be categorized?
- What would be the input into the city?
- What is the output from the city?
- What are some of the ways the city elements interact with each other?
- What is the role of electricity? Is it input, output, element, or boundary? Can it be more than one at once?

ASSESSMENT

Completion of System Parts Chart (Handout 2.1) by labeling parts of a city system.

System Parts Chart

1. What are the boundaries of the system? Why did you choose them? Were there other possibilities?

2. List some important elements of the system.

3. Describe input into the system. Where does it come from?

4. Describe output from the system. What part(s) of the system produce it?

5. Describe some important interactions:
   a. among system elements

   b. between system elements and input into the system

6. What would happen to the system if the interactions in 5a could not take place? In 5b?

Blueprints/Maps

LESSON LENGTH: Two sessions

INSTRUCTIONAL PURPOSE
- To introduce students to maps of the city that they are studying.
- To have students create a map of their selected area of the city.
- To help understand the role of electricity in their city.

MATERIALS AND HANDOUTS

Pencils
Markers
Bulletin board paper
Floor space

Maps
Building blueprints

Handout 3.1: Problem Log Questions
Handout 3.2: Problem Log Questions

Session 1

THINGS TO DO

1. Students work in groups, using city maps, papers, magazines, travel brochures, etc. to assist in mapping their section of the city (streets, buildings, etc.).

2. While students work, have them generate ideas about how other people will be able to use the map, if necessary. How will other people be able to interpret the symbols that the group is using? Discuss the procedure for labeling maps and creating map keys. Discuss what would happen if each person in a group or each group used different symbols on their maps.

3. Help students relate their area to the larger system (the city itself).

4. Now have students look at blueprints. Compare and contrast them with the map.

5. Discuss and illustrate idea of "bird's-eye" or "aerial view." Have groups develop drawing illustrating two examples of a bird's-eye or aerial view.

6. Have students create a map of a section of the city or building. This will serve as the beginning for creating a model and eventually wiring it for electricity.

THINGS TO ASK

- Think about other maps you've seen. How can you tell what the symbols on the map mean?

- How is your area of the city part of a system?

- How will you ensure that others will be able to read your map?

- How will a city official who needs to read all of the maps find common features?

- How are maps different from or similar to blueprints?

- Did working on the map of your area change your idea of the things you will have to think about as you create the electrical system? What new things did you notice?

- How did working on the map change your idea of the problem you face?

NOTE TO TEACHER

Students should be able to make map keys using symbols. Maps are most useful for the purpose of this activity when drawn from an aerial view. Some groups may also want to draw more detailed blueprints of buildings.

Session 2

THINGS TO DO

1. Return to the "Need to Know" board and see if any new questions have been raised from working with the maps. Some examples of likely questions are:

   - How is power delivered to different areas of the city?
   - Cities are provided with telephone, cable, gas, water, and sewage services (among others). How are these services delivered to consumers?
   - What is a power line and where are they located?
   - Who decides where they should be located? Why?
   - How do individual houses/businesses get electricity from the power lines?
   - How is electricity delivered within a building?
   - What is the role of a fuse box or circuit breaker in a building?
   - What are safety issues to consider when dealing with power lines or electrical lines? Why?

2. Have students draw a map of one floor of their house or apartment. A visit from a person from a local power company to discuss electrical hazards is advisable here. This person might also field some of the questions on distribution of power to various parts of the city.

ASSESSMENT

1. Completion of student-created maps serves as the assessment.*
2. Problem log activity reflecting on the issues which emerged as a result of creating the map.

EXTENSION

A field trip can be arranged for groups to visit the area of their model, providing students with more detailed information. Students can videotape the trip or snap photographs to assist them when building their models.

An additional copy of each map will be necessary for subsequent lessons. Each member of the group should also be provided a copy for his/her problem log.

NOTE TO TEACHER

This is also an appropriate place to present medical treatment of shock, lightning strikes, electrocution, etc.

HANDOUT 3.1

PROBLEM LOG QUESTIONS

Looking at problems from different perspectives often brings new ideas, information or situations to light. What did working on the map and/or blueprint reveal to you? Did the problem change at all in your mind? How?

1. Think about the process of creating a map. After today's activity, what are three most important pieces of advice you would give to an inexperienced map maker? Explain your choices.

2. Different maps tell us different things. What do you need to know about electricity in order to use a map? What maps seem best for this task? Explain your response.

3. Respond to the statement "Blueprints are just another kind of map."

Electricity Know-How

LESSON LENGTH: Two–three sessions (depending on time students need to complete station work)

INSTRUCTIONAL PURPOSE

- To use hands-on activities to learn about electricity and circuit components.

MATERIALS AND HANDOUTS

Electrical components (see lists in station handouts)
Books on electricity
Windows on Science laser disc (optional)
TV-laser disc player (optional)

Handout 4.1: Station One Handout
Handout 4.2: Station Two Handout
Handout 4.3: Station Three Handout
Handout 4.4: Station Four Handout
Handout 4.5: Station Five Handout
Handout 4.6: Problem Log Questions
Handout 4.7: Problem Log Questions (After Session 2)

NOTE TO TEACHER

If you already have modules or lessons on circuits and electricity, this would be an appropriate time to use them. One example is *Electric Circuits* from *Science and Technology for Children.*

Session 1

THINGS TO DO

1. Revisit the original problem statement (Handout 1.3). After reading the problem ask students, "What do we need to know about electricity in order to provide our city system with power (light circuits)?" Have students list the questions they want to have answered as they progress through the five electricity stations. Questions which cannot be answered at the stations should be assigned as independent research.

2. Set up five electricity stations. The five stations are designed to allow students to discover fundamental principles about electricity. Give each group a number from 1–5 to identify the first station they will visit. After 30 minutes at a station, rotate groups to the next station. Continue until each group has worked at each station.

THINGS TO ASK

- Is there any further information you need about electricity in order to power your particular section of the city?

- Do the power needs of the model you plan to build require additional knowledge about electricity? If so, what?

NOTE TO TEACHER

Depending on availability of materials, another format for learning general information about electricity may be substituted for the laser disc in Station Two. A text could be supplemented with other pictures and other resources.

Session 2

THINGS TO DO

1. Have students work in their groups to identify areas about electricity which are still unclear.

2. Give students the choice to revisit any station based on questions generated at the end of the first session. Materials can be placed in boxes or buckets at each station.

THINGS TO ASK

- Which areas about electricity are still unclear?

- Which station would you like to do again for clarification?

- What are your specific questions for this station?

- What will you do differently today to see if you can answer your questions?

- Which stations would you like to return to for new information?

- What are the new questions you are trying to answer?

- What will be your approach to get an answer to your question?

NOTE TO TEACHER

Teacher should demonstrate how to use wire stripper. Teacher's role is to monitor and guide students at stations, encouraging them to find answers to their questions through the literature at Station 4.

Session 3
After the Stations

THINGS TO DO

Ask students to return to their seats (in groups). Write a list of everything they have learned about electricity. Students share their lists with class. Update the "Need to Know" board in terms of what students have learned.

THINGS TO ASK

- How is a circuit constructed?
- How can you tell for sure that you have constructed a complete circuit?
- How does your understanding of circuits relate to your understanding of systems (studied in Lesson 2)?
- In one of the working circuits you observed, what are the boundaries, input, output, and elements?
- What questions from the "Need to Know" board about electricity are still unanswered?
- What new questions emerged as you worked?

NOTE TO TEACHER

Students should realize that the light bulb is the component that indicates a complete circuit has been made.

EXTENSION

Ask students to find out about the history of electricity. What role did Ben Franklin play? What were the contributions of Thomas Edison?

Electricity City

HANDOUT 4.1

STATION ONE HANDOUT

MATERIALS AND HANDOUTS

Insulated wire
Small flashlight bulbs
15 watt bulb
D-cell batteries
Wire cutters & strippers
Prediction Sheet One

STUDENT DIRECTIONS

Look at Prediction Sheet One and make a prediction for any six of the diagrams: Will the bulb light? After making your prediction, test it to find out if you were right. If you were wrong, why or why not? Briefly describe the part of your thinking that was mistaken.

Use the materials at this station to see if you can find other circuits that work. Mark the ones that do work on the sheet.

Can you add a circuit drawing to the prediction sheet for others to predict? (Remember, it does not necessarily have to "work"!)

PREDICTION SHEET ONE

Decide whether the bulb in each circuit will light. Write yes or no under each diagram.

HANDOUT 4.2

STATION TWO HANDOUT

MATERIALS AND HANDOUTS

Optical Data's *Windows on Science* laser disc segments on electricity
Laser disc player/monitor
Electricity program disc

STUDENT DIRECTIONS

Review the questions about electricity that you still need to have answered.

Watch the electricity segment of *Windows on Science.*

What did you learn here about electricity that you needed to know for your model?

Did this information raise any unanswered questions for you? Is there still information you need to know? Where or how might you find it?

HANDOUT 4.3

STATION THREE HANDOUT

MATERIALS AND HANDOUTS

Bulb holders
Flashlight bulb
Fahnestock clips
Different types of switches
D-cell battery holders
D-cell batteries
Insulated wire
Wire cutters & strippers
Prediction Sheet Two

STUDENT DIRECTIONS

Look at Prediction Sheet Two and make a prediction for any seven of the diagrams: Will the bulb light? Why or why not? After making your prediction, test it to find out if you were right. If your prediction was wrong, briefly describe the part of your thinking that was mistaken.

Play with the materials at this station and see if you can find other circuits that work.

Can you add a circuit drawing to the prediction sheet for others to predict? (Remember, it does not necessarily have to "work"!)

PREDICTION SHEET TWO

Decide whether the bulb in each circuit will light. Write yes or no under each diagram.

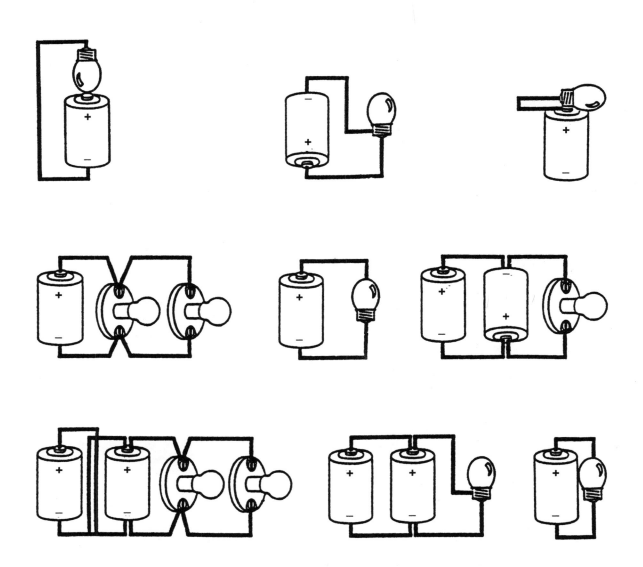

HANDOUT 4.4

STATION FOUR HANDOUT

MATERIALS AND HANDOUTS

Power company literature

Books, encyclopedias, pamphlets, brochures (containing information on electricity)

Clear light bulb

Light bulb with glass broken and removed

STUDENT DIRECTIONS

Review the list of things you need to know about electricity. Look for the answers to your questions in the literature available at this station.

What makes a light bulb actually light? Draw a diagram and explain what is happening. How did you find this out?

What is a resistor? How is the filament of the light bulb a resistor? What does this do for the bulb?

HANDOUT 4.5
STATION FIVE HANDOUT

MATERIALS AND HANDOUTS

One circuit tester (a simple circuit made of a D-cell battery in a holder, a bulb in a socket wired as follows):

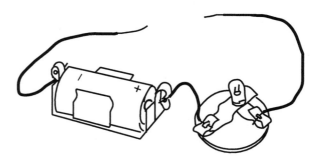

Materials to test such as:
- Aluminum wire
- Nail
- Marble
- Wooden pencil with no eraser
- Brass fastener
- Paper clip
- Styrofoam
- Plastic soda straw

STUDENT DIRECTIONS

Conductors are materials that allow electricity to pass through them (and thus completing the circuit). Insulators are materials through which electricity does not flow (at least not in amounts that are detectable).

Use the circuit tester and materials at this station to determine whether each is either a conductor or insulator. Make a list of which materials are conductors and a list of which materials are insulators.

HANDOUT 4.6

PROBLEM LOG QUESTIONS

1. Think about what you have learned about electricity in each of the five stations. Draw a small web of the ideas associated with each station. How do these ideas fit together? Create a web to indicate how these ideas about electricity fit together!

2. After your web is complete, look at the original list of questions about electricity. Can you locate on your web where the information is to answer the original questions? Are there still questions unanswered? Where can you go to find the answers?

3. How does "electricity" travel?

HANDOUT 4.7

PROBLEM LOG QUESTIONS (AFTER SESSION 2)

Most of you had to return to a station today to clarify some issues. What could you have done the first time to avoid having to return? What corrections did you make in your procedure today?

Circuit Experiments

LESSON LENGTH: Two sessions

INSTRUCTIONAL PURPOSE

- To design and conduct an experiment on various components of an electrical circuit.

MATERIALS AND HANDOUTS

Components of a circuit (bulbs, insulated wire, Fahnestock clips, batteries, etc.)

Handout 5.1: Student Brainstorming Worksheet
Handout 5.2: Student Experiment Worksheet
Handout 5.3: Student Protocol Worksheet
Handout 5.4: Laboratory Report Form
Handout 5.5: Problem Log Questions

Session 1

THINGS TO DO

1. Look at the "Need to Know" board and see if there are any unanswered questions about electrical circuits that could be answered by doing an experiment. If not, prompt students with questions such as, "What do you think would happen if you put two bulbs in your circuit instead of one?" and "How could you find out?"

2. Show students the available materials. Have students generate a list of unanswered questions that could be answered by doing an experiment with some of the materials. Examples include:

   - What is the effect of the length of the wire on the brightness of the bulb?
   - What is the effect of the number of batteries in a circuit on the brightness of the bulb?

- What is the effect of the number of bulbs in a circuit on the brightness of the bulb?
- What is the effect of the number of bulbs in a circuit on the life of a battery?
- What is the effect of using a different size battery (D-cell, C-cell, 9 volt) on the brightness of the bulb?

3. Give students the Student Brainstorming Worksheet (Handout 5.1) to complete in small groups. Each group may choose a different experimental question or the whole class may address the same one.

4. Have students design an experiment to answer their question. They should use the Student Experiment Worksheet (Handout 5.2) and the Student Protocol Worksheet (Handout 5.3) as a guide.

5. Review student designs for experiments before the next session.

THINGS TO ASK

- Why is it important to know the answer to your experimental question?
- How many trials (repetitions of the experiment) should you do? Why?
- What things can you do to make the results of your experiment reliable?

Session 2

THINGS TO DO

1. Have students carry out their experiments.
2. Have students complete the Laboratory Report Form (Handout 5.4).
3. Discuss the results of the experiments. Modify the "Need to Know" board as needed.

THINGS TO ASK

- What new questions does your experiment raise?
- How sure are you of your results? Explain.
- If you did the experiment over, what would you do differently?
- How could your experiment be considered a system?

ASSESSMENT

All student worksheets may be used to assess individual and group progress.

EXTENSION

Have students complete another experiment from the list of possibilities generated in class and report results to the class.

NOTES TO TEACHER

1. If your students have not had experience in designing experiments, they will need additional work in this lesson. You may want to consult the following resources:

   Cothron, J.H., Giese, R.N., & Rezba, R.J. (1996). *Science experiments and projects for students*. Dubuque, IA: Kendall/Hunt Publishing Company.

   Cothron, J.H., Giese, R.N., & Rezba, R.J. (1996). *Science experiments by the hundreds*. Dubuque, IA: Kendall/Hunt Publishing Company.

   Cothron, J.H., Giese, R.N., & Rezba, R.J. (1996). *Students and research: Practical strategies for science classrooms and competition*. Dubuque, IA: Kendall/Hunt Publishing Company.

2. Students may need to construct a "brightness meter" for calibrating brightness of the bulb in some of the experiments. This can be done by stapling overlapping strips of paper of various lengths and then holding it over the bulb until the glow can be seen. (Section 1 will have one layer of paper, Section 2 will have two layers of paper, and so on.)

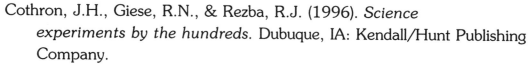

| Layers of paper | 1 | 2 | 3 | 4 | 5 | 6 | 7 | 8 | 9 |

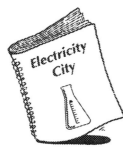

HANDOUT 5.1

STUDENT BRAINSTORMING WORKSHEET

1. What do we need to find out? (What is the scientific problem?)

2. What materials do we have available?

3. How can we use these materials to help us find out?

4. What do we think will happen? (What is our hypothesis?)

5. What will we need to observe or measure in order to find out the answer to our scientific question?

Adapted from: Cothron, J. G., Giese, R. N., & Rezba, R. J. (1989). *Students and research*. Dubuque, IA: Kendall/Hunt Publishing Co.

STUDENT EXPERIMENT WORKSHEET

Title of Experiment:

Hypothesis (Educated guess about what will happen):

Independent Variable (The variable that **you change**):

Dependent Variable (The variable that responds to changes in the independent variable):

Observations/Measurements to Make:

Constants (All the things or factors that remain the same):

Control (The standard for comparing experimental effects):

1. List the materials you will need.

2. Write a step-by-step description of what you will do (like a recipe!). List every action you will take during the experiment.

3. What data will you be collecting?

4. Design a data table to collect and analyze your information.

HANDOUT 5.4

LABORATORY REPORT FORM

1. What did you do or test? (Include your experiment title.)

2. How did you do it? What materials and methods did you use?

3. What did you find out? (Include a data summary and the explanation of its meaning.)

4. What did you learn from your experiment?

5. What further questions do you now have?

6. Does the information you learned help with the problem?

Handout 5.5

Problem Log Questions

1. What did you learn from your experiment that will be useful in wiring your model? Explain why.

2. Draw a picture of your experimental setup. Label its boundaries and its elements. List the input and put into your experimental system and the output that came out of it. What interactions inside the system allowed it to produce output? Were there interactions between the original system elements or interactions with input that you added?

6

Generating Power

LESSON LENGTH: Three sessions

INSTRUCTIONAL PURPOSE

- To determine the differences between AC (alternating current) and DC (direct current).
- To explore how a power station produces electrical power.
- To develop an understanding of power outages, black-outs, brown-outs, and surges.

MATERIALS AND HANDOUTS

One red ball or marble for each student

Copy of newspaper or magazine articles concerning the power outages across Arizona, California, Colorado, Idaho, Nevada, Oregon, Utah, Wyoming, Alberta (Canada), and British Columbia (Canada) on July 2, 1996

Various materials concerning electrical power production, power outages, black-outs, brown-outs, and surges

Handout 6.1: AC/DC Know-How

Handout 6.2: AC/DC Activity

Handout 6.3: Simple Generators

Handout 6.4: AC/DC Activity Questions

Handout 6.5: Problem Log Questions

Handout 6.6: Power Outage Questions

Handout 6.7: Problem Log Questions

Session 1

THINGS TO DO

1. Review the material covered in Lesson 4. Refer to the "Need to Know" board and discuss what particular electrical phenomenon students still need to know more about.

2. Break the class into small groups and distribute Handout 6.1. Have each student read the section and research the information needed to answer the questions. If possible, have some resources available in the classroom for this purpose. Groups should come to some consensus about the discussion questions.

3. Bring the class back together to review the questions as a whole, referring to the "Need to Know" board when appropriate.

4. Distribute Handout 6.2 and do the AC/DC activity.

5. Prompt the discussion to move to the topic of power plants and how they generate energy, by giving out Handout 6.3. Stress the same process no matter what type of plant (hydroelectric, nuclear, coal, geothermal, etc.) is being considered.

6. Have students complete Handout 6.4 (AC/DC Activity Questions)

7. Have students respond to Problem Log Questions (Handout 6.5).

THINGS TO ASK

- How has the "Need to Know" board changed from what we learned yesterday?
- What do we still need to find out?
- How does alternating current fit into our problem?
- What are the differences between AC and DC current?
- How do these differences impact our problem?
- How do power plants produce electricity?
- Is there a common way to make AC current?
- How are the types of power plants similar?
- How are they different?

Session 2

THINGS TO DO

1. Distribute the newspaper report or magazine article on the severe power outage on July 2, 1996. Have all students read the article thoroughly.

2. Break the class into small groups and distribute Handout 6.6 (Power Outage Questions). Allow students to use various materials on electrical power as well as the article to discover the answers to the questions on Handout 6.6.

3. Once students have had adequate time to come up with answers supported by facts, bring the class back together again. Refer to the "Need to Know" board in light of the information the students have discovered.

4. Have students answer the Problem Log Questions (Handout 6.7).

THINGS TO ASK

- Does the power outage have any effect on our immediate problem?

- How do you think this type of power outage can occur?

- Based on what we know about power generation, could we come up with a theory on power outages like this?

- What is the relationship of black-outs and surges to our problem?

- What do we still need to know and how can we find it out?

Session 3

THINGS TO DO

1. Have groups or teams of students select from the following topics: medical treatment of electrical shock, lightning strikes, electrocution, proper insulation techniques, medical uses of electricity (i.e., heart failure, resuscitation).

2. Have student groups research their selected topic. Provide resource materials, trip to the library, or other sources (computers, class visitors, etc.).

3. Each group should be prepared to present information on their topic to the entire class.

THINGS TO ASK

- How can you go about getting quick, reliable information on your topic?

- How do you know an information source is reliable?

- What implications does your group's topic have on our overall problem?

ASSESSMENT

1. Problem Log Questions.
2. Participation in small group discussions and research.

EXTENSION

1. Learn about watts/kilowatts, kilowatt-hours, and how to read the dial on an electric meter.

2. Find out about the cost of electricity in your area. How does this compare to other areas of the country?

3. Give students a set of real data (kilowatt-hours used) from a set of electric bills from a residence (either house or apartment). Have them make a graph of the number of kilowatts used over the course of the year. What could account for fluctuations in use?

Electrons moving through a wire can move continuously in the same direction, or they can change direction back and forth over and over again. When electrons always flow in the same direction, the current is called direct current (DC). Electricity from dry cells and batteries is direct current. When electrons move back and forth, reversing their direction regularly, the current is called alternating current, or AC. The electricity in your home is alternating current. In fact, the current in your home is changing directions 120 times every second. Although direct current serves many purposes, alternating current is better for transporting the huge amounts of electricity required to meet people's needs.

1. Why would a power distribution system (i.e., a power plant) be less effective without alternating currents?

2. In direct current circuits, electrons actually move through the current in the same direction. How can power move through an AC circuit if the electrons keep changing direction?

3. Do the electrons actually need to "go" anyplace to transmit power?

Handout 6.2
AC/DC Activity

Procedure

1. Each student sits down on the floor, facing the center and holds a red ball or some other appropriate object.

2. The students represent a circuit and the red balls represent the electrons.

3. When the teacher says "switch on," all students pass their red balls to the right. Students should count the number of times they passed an "electron." They keep passing "electrons" until the teacher says, "switch off." This is a model for direct current.

4. Now we are going to repeat the process, but add one more step. The teacher is going to say "switch on" and students will begin to pass "electrons." Every 5 seconds or so, the teacher is going to say "change directions" and instead of passing the electrons to the right, begin passing them to the left. Each student should keep count of the total number of electrons that they pass to other students, no matter which direction they were passed. At some point, the teacher will say "switch off." This is a model for alternating current.

5. Each student should complete Handout 6.4.

HANDOUT 6.3

SIMPLE GENERATORS

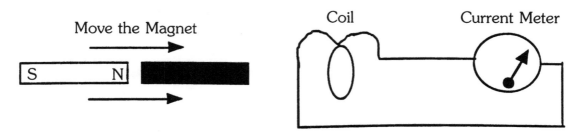

The simple current generator above works by moving a magnet near a coil of wire producing an electric current in the wire, indicated by the current meter. The motion of the magnet is important to creating the current. Merely holding the magnet near the coil of wire does not generate electricity.

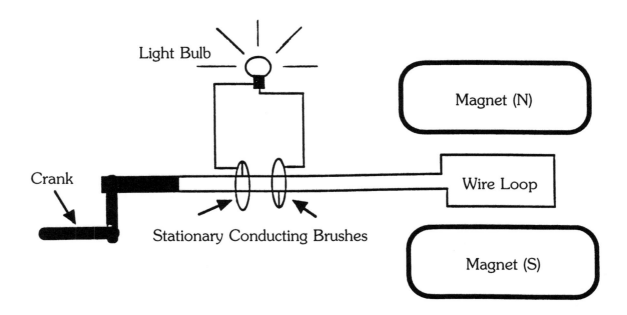

In a simple alternating current generator, a wire loop rotates between magnetic poles, giving rise to an electric current. Each time the wire loop flips, the current changes direction. Stationary conducting brushes contacting the circular rings convey the current from the rotating loop to wires that carry the current to the light bulb.

HANDOUT 6.4
AC/DC ACTIVITY QUESTIONS

1. Imagine what might happen if you doubled the number of balls you had to handle. Would it be easier or harder to pass on electrons? What if you had only half the number of balls?

2. Now imagine if you suddenly had ten times the number of balls to transfer. Would you be able to do this?

3. What conclusions can you draw between this activity and the strain on power generators? Explain your reasoning.

Now that you know about AC/DC and how power plants produce electricity, how is this going to affect our problem? What issues are yet unresolved about power generation and the problem? Explain your responses.

HANDOUT 6.6

POWER OUTAGE QUESTIONS

1. What factors about power generation could cause power stations to shut down one after another like dominoes?

2. What is a black-out and a brown-out and how do they differ?

3. What are power surges? Why are they dangerous?

4. If two generators are running to meet the power needs of a community and one of them suddenly shuts down, what happens to the other?

5. What would happen to a power generator if everyone in a community turned their lights on at the same time? How about if they all turned them off?

HANDOUT 6.7

PROBLEM LOG QUESTIONS

1. Now that we know about power outages and their effects on power generation, how are we going to incorporate this information into our problem?

2. When we design the retirement complex, what steps can we take to insure a safe supply of power gets to our (retirement) complex? How can we protect against black-outs, brown-outs, and surges?

Consultation with an Electrical Expert

LESSON LENGTH: Three sessions

INSTRUCTIONAL PURPOSE

- To get information about electricity and power generation through interaction with an electrical expert, such as an electrician, a lineman, or a representative from the local power company.

MATERIALS AND HANDOUTS

Chart and markers

Audio-Visual equipment for guest speaker

Handout 7.1: Visitor Planning Sheet

Session 1
Before the Speaker Comes

THINGS TO DO

1. Brainstorm with students, deciding what questions need to be asked of the speaker. Use the "Need to Know" board to choose questions.

2. Class discussion can help sort questions into most and least important questions.

3. Students should also be guided to think about the best way to phrase the questions. Are they specific enough? Are they too specific?

4. Group questions can be recorded on a master question chart.

5. Students can then add any of their own questions to individual Visitor Planning Sheets (Handout 7.1).

THINGS TO ASK

- What information do we want to know?

- What information will the guest speaker be most qualified to give?

- What do we want to know by the time the guest speaker leaves?

- What facts do we want to get from this person?

- What opinions would be interesting to have?

- Which of these questions are most important?

- How can we get an idea of this person's perspective on this kind of situation?

- Do you think this person will have a bias? What would it be? How can we find out?

Session 2
The Guest Speaker's Presentation

THINGS TO DO

1. Guest Speaker: The guest provides his/her information regarding the area of his/her expertise.

2. Students take notes and ask their questions.

3. Students should also be prepared to share with the guest speaker background on the problem and their decisions to date.

Session 3
Debriefing

THINGS TO DO

1. In a follow-up to the guest speaker, teacher and students should review the "Need to Know" board, removing questions which have been answered and adding new issues, if necessary.

2. Teachers and students should discuss the potential bias in the information provided by the guest speaker and the possible effects of that bias on the validity of the information.

THINGS TO ASK

- What were the things we learned from the guest speaker?

- How does the new information affect our thinking about the problem?

- Do we need to reorganize our approach to the problem?

- Did this person reveal a particular bias? If so, what?

- Where can we go to get another perspective? A balanced report of information?

ASSESSMENT

1. Students should report in their problem logs information provided by the guest lecturer and reflect on the potential of bias in the problem log.

2. Students write a thank-you letter to the guest speaker, detailing which information was particularly helpful.

NOTE TO TEACHER

If the expert comes to the classroom, all students can participate. This format can also be used by small groups who need to interview an outside expert outside of class; afterwards, they can report any new information to the class.

HANDOUT 7.1

VISITOR PLANNING SHEET

Student Name _____

Name of Visitor _____

Who is this visitor?

_____

_____

_____

Why is this visitor coming to see us?

_____

_____

Why is this visitor important to us?

_____

_____

_____

What would you like to tell our visitor about our problem?

_____

_____

What questions do you want to ask the visitor?

_____

_____

_____

Models and Scale

LESSON LENGTH: One session

INSTRUCTIONAL PURPOSE

- To plan the construction of students' model city or building, taking the concepts of scale and proportion into account.

MATERIALS AND HANDOUTS

Small Matchbox cars and other examples of models

One piece of black construction paper

Handout 8.1: Group Activity
Handout 8.2: Problem Log Questions
Handout 8.3: Problem Log Questions

NOTE TO TEACHER

Materials lists will vary depending on group needs. Sometimes it is better to use a building as your replication model so it is easier to measure and design.

THINGS TO DO

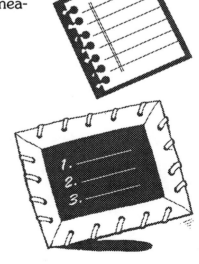

1. Discuss with the important features of a model. Ask students to brainstorm a list of toys that are models of adult objects. Ask them whether or not items in the model have proportional relationships. Speculate about the consequences of having a model where components are not proportionate. Ask students how they would go about designing a proportional model. Introduce the concept of scale.

2. Conduct a demonstration of the concept of proportion. Hold a Matchbox car next to a piece of black construction paper. Ask a student to come to the front of the room and cut the construction paper into the size and shape of a road that would be the correct size in relationship to the car.

3. Discuss the advantages of using a common point of reference (such as the car) when determining the proportions in a scaled model. Ask students to decide whether or not all of the models should use a common point of reference as they develop their models. Ask students to select the point of reference to use for scaling their model.

4. Have students in their groups or teams complete Handout 8.1.

5. Discuss as a class each group's checklist and why particular items were selected.

6. Have students complete Problem Log Questions (Handouts 8.2 and 8.3).

THINGS TO ASK

- Look at some examples of models we have in the room. What would you say are some of the characteristics of the models?

- What are some models you have at home? Do those models share the same characteristics?

- If you saw a car model in the store that had wheels that were twice as large as it should would you buy it? Would you consider it a good model? Why or why not?

- Think about the model you are about to build. Should your model meet all of the criteria for a good model that you just listed?

- What scale do you recommend we use? Why?

- What would happen if we used a toy car to determine proportion for the road and then used a brick to determine the proportion of the mailboxes?

- Should all of the models use the same scale, or are different scales OK?

- What would be important to consider as you make that decision?

- Now that you have determined your scale, create a list of materials you will need to build your part of the city or building. What are the materials which would be proportionate (or could be made proportionate) to your point of reference?

ASSESSMENT

1. Students evaluate themselves using a checklist. Students compare their maps to the materials needed to construct a three dimensional model.

2. Problem Log entry reflecting on the nature of planning and proportion as parts of this system.

EXTENSION

Instead of estimating size, students can use other math skills and actually measure roads, building, and scale the size down to the model size (e.g., 1" = 10 ft., etc.).

HANDOUT 8.1

GROUP ACTIVITY

1. Think of all the ways your map (from Lesson 3) can help you build your model of the city/building.

2. List all of the parts of a good model. Next, list some ways that you could "test" the model to see if it's good. Create a "model making" checklist using the items on your list. USE THIS CHECKLIST as you develop your model.

3. Make a list of all of the materials you are going to need to build your model.

HANDOUT 8.2

PROBLEM LOG QUESTIONS

We have talked about "systems" already in this unit. What part does proportion play in the creation of the system you are working on? Is "proportion" the same thing as "scale"? How are they the same/different?

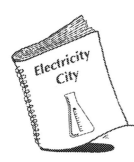

Handout 8.3

Problem Log Questions

You have already thought some about your model. Did today's exercise change your idea about your model? What has your path to creating your model taught you about planning? What would you say are some of the keys to good planning?

lesson

9

Wiring Diagrams and Circuitry

LESSON LENGTH: Two sessions

INSTRUCTIONAL PURPOSE

- To help students develop wiring diagrams for their model section of the city.
- To help students plan the construction of the circuits necessary for lighting their model city.

MATERIALS AND HANDOUTS

Foam core display board

Shoe boxes

Small boxes

Green tissue paper

Black background paper

Manila construction paper

Multicolored construction paper

Kleenex boxes

Large bags or boxes for storage (one per team)

D-cell batteries & battery holders

Wire cutters & stripper

Different types of switches

Flashlight bulbs or small sections of Christmas tree lights (5 bulbs) with wire stripped at ends

Bulb holders

Fahnestock clips

9-volt batteries

Handout 9.1: Problem Log Questions

Handout 9.2: Materials Planning Sheet

Handout 9.3: Problem Log Questions

Session 1

THINGS TO DO

1. Present students with electrical symbols to use when planning their models (a good resource is *Electric Circuits, Science and Technology for Children*. Washington, D.C.: National Science Resources Center, 1991).

2. Give students a ruler, paper, and a pencil. Have student draw an original plan of the model city or building they have chosen.

3. Instruct students to plan where power is needed in their model and to draw their original plan of the model. In a tutorial discussion, have students identify the criteria they will need to consider as they plan the circuitry for the complex. Make sure that students state their assumptions about the electrical needs of their selected section of the city/building.

4. Using their plan for their model city or building, have student draw an electrical plan for their model. Have them label with a red arrow the flow of the current (from positive terminal of battery to negative). Have them label with a blue arrow the flow of the electrons (from negative terminal of battery to positive). Ben Franklin defined the direction of current backwards relative to the flow of electrons.

5. Have students complete Problem Log Questions (Handout 9.1). Have groups that exchange plans discuss their recommendations to the original group.

THINGS TO ASK

- What are the general electrical needs of your part of the city/building?

- How can you indicate that on a schematic drawing?

- Are the needs consistent across the section or do different sections have different needs?

- What are the specific needs? What are the general needs?

- How will you accommodate all of these needs?

- How do the sub-systems interact?

- Will the systems interactions inhibit one another? How can you avoid that?

Session 2

THINGS TO DO

1. Assist students as they use their model maps and electrical plans to determine the amounts of various materials for the model (Handout 9.2). Remind students that as electrical emergency assistants, part of their assessment will include a good planning of materials (they should not create overly generous materials lists to avoid thinking about the task).

2. In a large group discussion, review the checklists students previously developed for judging the needs of each area of their model. Discuss the process of material planning with students.

3. Review the materials lists developed by students group by group. Help the student identify the materials which they can find and those which the teacher will have to locate. Help students develop a list of resources they can check to find the location and prices of materials they cannot supply.

4. Have students complete Problem Log Questions (Handout 9.3).

THINGS TO ASK

- Look at the different areas of your model. What challenges would you anticipate based on the physical location of the properties?

- What special accommodations might you have to make for special conditions?

- How are high population and low population areas different in their demands for electricity?

- How could you tell the length of wire needed to reach each part of the city?

- How does the capacity of the batteries affect your design? How many bulbs can you light from the different kinds of batteries you have used so far?

- How can you maximize efficiency while minimizing costs? What elements of your electrical design would you change to meet these criteria?

ASSESSMENT

1. Complete electrical plan serves as the evaluation.
2. Students self-assess their own lists against the criteria checklist they developed.
3. Teacher assesses both the materials list and the self-assessments.

NOTE TO TEACHER

Students may use city maps to help them develop their electrical plan. Materials listed for construction of the models may vary. Pilot teachers found that modification to already constructed boxes (i.e., shoe boxes, Kleenex boxes, etc.) worked well, even if some sides had to be cut and refitted to accommodate the scale being used. These materials are provided so that students can determine what to use, what is available, and what remains to be procured. Providing some kind of large bag or box to each team as storage for materials is helpful.

The following is a basic set of electrical symbols that can be used for drawing circuit diagrams.

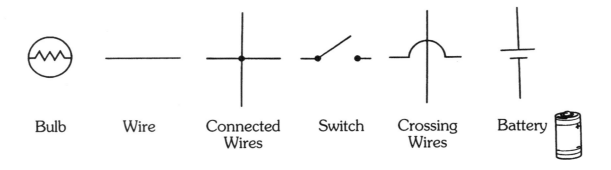

| Bulb | Wire | Connected Wires | Switch | Crossing Wires | Battery |

HANDOUT 9.1

PROBLEM LOG QUESTIONS

1. Does your electrical plan include all the electricity you want in your model? If not, what did you leave out? Why?

2. What are the symbols in your key to explain the parts of your electrical plan?

3. Is everything clear and well labeled? Could someone else read and interpret your plan? What makes the plan especially clear?

4. SWITCH PLANS WITH ANOTHER GROUP! Read their plan and describe it. Could you wire their system based on their plan? Why? Why not? Make a list of recommendations and return the plan to the original group.

Handout 9.2

Materials Planning Sheet

1. What are the electrical needs of your model (in other words, for what are you actually using electricity)? Remember to check your electrical plan!

2. In order to meet those needs, how much of which materials will you need?

    Wire?

    Bulbs?

    Bulb Holders?

    Generator?

    Power Source(s)?

    Clips?

    Switches?

    What Else?

3. How did you decide how much of these materials you will need?

4. What are the uses for electricity in your model? How is this different from electrical needs in real life? What else should you consider if you were doing actual repairs in a city hit by an emergency situation?

Handout 9.3
Problem Log Questions

In today's discussion, we included two new ideas for you to think about as you plan your electrical system: location and capacity. How did these criteria alter your idea of the problems you might face while constructing your model?

Model Construction

LESSON LENGTH: Multiple sessions

INSTRUCTIONAL PURPOSE
- To facilitate the actual construction of student models.
- To encourage reflective thinking and metacognition while models are under construction.

MATERIALS AND HANDOUTS

Building materials selected by students in previous lessons

Handout 10.1: Electrical System Progress

NOTE TO TEACHER

Prior to the start of the lesson, an excellent opportunity exists to have an architect visit the classroom to talk to the students about model building. A similar visitor format from Lesson 6 may be used.

THINGS TO DO

1. Provide students with materials to construct their model and electrical systems within their model.

2. Facilitate as the groups work on wiring and lighting their portion of the city or building. Have students use Handout 10.1 as a daily monitoring device. Because the wires break loose easily, it is best not to glue things into place until the model is finished. Students should draw a schematic first. Then they should place the circuit on the board. When everything is in place, glue it. Then tape the wires with electrical tape. (**Note:** Depending on the complexity of student models, the length of this lesson will vary greatly. It may take up to 10 hours of classtime.)

3. Have students actually light the model. If it doesn't light up, then have students investigate and correct their lighting problems.

THINGS TO ASK

- How many bulbs can you light from a 9 volt battery?
- What is the most efficient way for you to wire your model?
- What parts of your model city system are affected by your electrical plan?
- Does your electrical plan handle the needs of your section of the city or building?

ASSESSMENT

1. Completion of a model with accurate proportions.
2. Completion of the wiring system.
3. Analysis and correction of problem areas.
4. Cooperative efforts among team members.
5. Completion of Electrical System Progress Worksheet (Handout 10.1).
   (Teacher to determine how often copies of this sheet need to be completed by the student/team.)

Date: _____

Location of Electrical System: _____

Group: _____

1. What did you accomplish today with your electrical system?

2. What problems are you having?

3. How are you planning to solve those problems?

Lesson

11

Power Outages

LESSON LENGTH: One session

INSTRUCTIONAL PURPOSE

- To allow students to analyze the role of the electrical power system as an element of a larger system, namely the city.
- To help students to identify the other elements of the city system with which the electrical power system interacts.

MATERIALS AND HANDOUTS

Handout 11.1: Compounding the Problem
Handout 11.2: Primary and Secondary Effects Chart
Handout 11.3: Problem Log Questions

THINGS TO DO

1. Review with students the structure of their system and its components. Ask students what kinds of interactions they discovered at work among the various pieces of their system.

2. Discuss with students the role of the electrical system as a part of the larger system that is the city: what does the electrical system affect? Introduce the idea that, as an element of the larger system, the electrical system affects functioning of other subsystems in the city. Have students list other city subsystems.

3. Discuss the idea of primary and secondary effects. Use the example of the storm, discussed in the Additional Problem Statement, to brainstorm a list of the potential primary and secondary effects of power outages. (Example: Primary—knocking down powerlines; secondary—lights go out, someone gets caught in an elevator.)

4. List the other city subsystems and the potential consequences for each subsystem of a power outage.

5. Tell students that they are to create a disruption in their system. Each disruption plan should include a description of the potential secondary effects the disruption could have on other subsystems which interact with the electrical system. Each disruption plan should meet specific criteria including: effect on human life/safety; time constraints for fixing the problem; realism; and identifiable source.

6. Provide students with time in class to develop their "compounded problem" scenarios.

7. Have students complete Handouts 11.1–11.3.

ASSESSMENT

1. Completion of "compounded problem" scenarios.

2. Problem Log Handout.

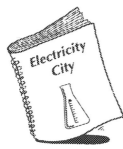

HANDOUT 11.1

COMPOUNDING THE PROBLEM

1. Create a "compounded problem" which includes the setting for the power outage (your model site) and what happens when the power goes out.

2. Indicate any other systems (or parts of systems) that might be affected by your "compounded problem" using the primary and secondary effects chart.

3. What happens at your house when the power goes off? What are some things that are different in your home when the power goes off?

4. What backup facilities are available in your school building in case the power goes out on a school day? Why do exit lights stay on when the power goes out?

Electricity City

HANDOUT 11.2

PRIMARY AND SECONDARY EFFECTS CHART

Your power outage will directly impact some things and only indirectly impact other things. Therefore there will be both primary and secondary effects of your problem on the community.

Use this chart to predict the ways your power outage will directly impact things (primary impacts), and the way changes in the primary impact also changes other things (secondary impacts).

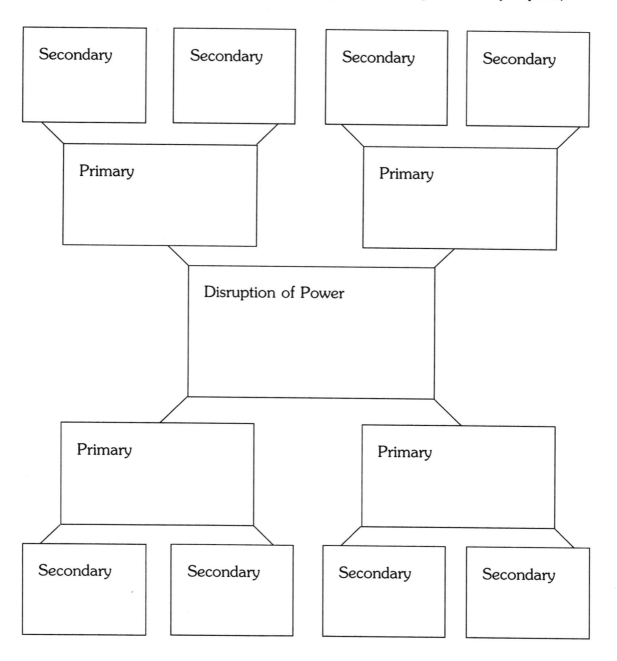

Secondary Secondary   Secondary Secondary

Primary   Primary

Disruption of Power

Primary   Primary

Secondary Secondary   Secondary Secondary

HANDOUT 11.3

PROBLEM LOG QUESTIONS

What did you learn about problems as a result of this activity? What makes a problem problematic? When does a situation turn into a problem?

The Storm Hits

LESSON LENGTH: One or two sessions

INSTRUCTIONAL PURPOSE

- To provide students with the skills to realistically predict potential damage to their systems resulting from a particular electrical system failure.

MATERIALS AND HANDOUTS

Wire clippers
Copies of electrical drawings from Lesson 8
Constructed models

THINGS TO DO

1. Guide a discussion to help students identify the parts of their model that would change if their "compounded problem" (from Lesson 10) actually happened.

2. Prior to the "storm," have one student in the group code a copy of their electrical drawing (completed previously) to indicate the faults in the model. Then allow students time to "de-construct" their model according to their "compounded problem," clipping wires, turning over light poles and trees as necessary to disconnect or otherwise break the existing circuits.

THINGS TO ASK

- What would happen to different parts of your model if your problem really happened?

- What could happen to natural parts of the system?

- What could happen to man-made parts of the system?
- What are the constraints on the damage caused? Or, could absolutely anything happen?

ASSESSMENT

Faults indicated on the electrical drawing should match faults on the model. Students evaluate for thoroughness. Teacher assesses for realism and for accuracy of self-assessment.

Repairs

LESSON LENGTH: One or two sessions

INSTRUCTIONAL PURPOSE

- To help students transfer understanding of circuits and wiring to reconstruct another group's model.

MATERIALS AND HANDOUTS

Partially dismantled models (from Lesson 12)

Written account of compounded problem

Blank copies of the electrical drawings of each model

Handout 13.1: Scientific Reasoning Sheet

Handout 13.2: Problem Log Questions

THINGS TO DO

1. Assign each deconstructed model at random to a different group to solve. Students should both fix the electrical system(s) and account for the secondary effects of the problem as well. They should record repairs on a copy of the electrical plan.

2. Students can be given a time limit for fixing the problems. This deadline adds excitement to the problem-solving process.

3. During the problem-solving activity, groups should have a recorder complete the Scientific Reasoning Sheet (Handout 13.1) to reflect how they worked through the problem.

4. Have students complete Handout 13.2. Sharing these answers as a class activity is an excellent learning opportunity.

THINGS TO ASK

- What is the most important part of this problem?
- What are the different parts of the model which need to be fixed?
- What are the different approaches you are considering?
- What are the advantages/disadvantages of your approach?

ASSESSMENT

The solving of the problems successfully serves as one type of overall unit assessment. Criteria for judgment are:

- Students' ability to identify on the blank electrical drawing the places where components of the system had to be fixed.
- Realism of the resolution of the secondary effects.
- Successful repairs of circuits.

HANDOUT 13.1

SCIENTIFIC REASONING SHEET

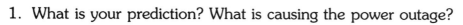

1. What is your prediction? What is causing the power outage?

2. What are you going to do?

3. Was your prediction confirmed? If not, what's your next step?

HANDOUT 13.2
PROBLEM LOG QUESTIONS

What have been the most successful approaches you have used to solve the problem?

lesson

Back to the Original Problem: Design of the Complex

LESSON LENGTH: Two–five sessions

INSTRUCTIONAL PURPOSE

- To help students identify thinking strategies which facilitated their design and repair of their models.
- To help students transfer these skills to the design of the recreational complex.

MATERIALS AND HANDOUTS

Paper (large size)

Rulers (an architects rule can be used if students are familiar with this tool)

Pens, pencils, and markers

Various sizes of drafting squares, triangles, and associated templates

Session 1

THINGS TO DO

1. Have students read from their problem logs their ideas about the strategies they found to be most effective and the most difficult problems they faced while model building and wiring (Handout 13.2). Develop a list of the successful strategies and most difficult problems.

2. Ask students to reflect on the strategies and techniques they used to solve the sub-problem. Discuss the difficulties they encountered and the specific strategies they used (or should have used) to get beyond the difficulties.

THINGS TO ASK

- What strategies did you learn from previous activities that will help you with designing a complex?

Sessions 2–5

THINGS TO DO

1. Have students discuss their ideas for the complex together in their team. Have students list components of the complex and certain aspects of construction and electric power supply that they want to incorporate in their design.

2. Have teams draw a plan of their design. Remind them that they are a representative of the power company and that the electrical wiring of the complex is a major part of their task as a design team member.

THINGS TO ASK

- What important features of the building and circuitry do you need to be sure to consider in your design?

- What other important features should your design include to be useful to its intended population?

- Is your design practical? Will the city consider buying it?

- Does your blueprint incorporate all the necessary symbols so that others can read it?

NOTE TO TEACHER

Specific requirements or constraints (such as size of complex, height, specific population requirements, etc.) should be developed to give students a base to operate from, much like city planners give architect firms. The particular requirements should coincide with the type of complex used in the Problem Statement (i.e., retirement community might need a certain number of apartments or a sports complex may need a pool and open gym with 100 lockers). If your school has a computer-aided design program, this is an excellent opportunity for students to utilize the technology.

Presentation of the Complex Design

LESSON LENGTH: Two–five sessions

INSTRUCTIONAL PURPOSE

- To give students an opportunity to synthesize and integrate the information they have learned throughout the unit.

MATERIALS AND HANDOUTS

Poster board, chart paper, graph paper, markers, audio-visual equipment, chart stand, and podium, etc. that students might need to present their complex design

Handout 15.1: Presentation Planning Sheet
Handout 15.2: Persuasive Presentation Assessment Form
Handout 15.3: Problem Log Questions

Sessions 1–3

THINGS TO DO

1. Inform students that their architectural team has been asked to present their complex design before City Council and other important community members.

2. Have students (in their teams) discuss a method of presentation for their design. They can be encouraged to develop reports, displays, graphs, charts, pictures, etc. Be sure to stress that this presentation is important to their company—if they get the contract, it means more business for their company. Therefore, their presentation must "sell" the design to the City Council. Have students complete Handout 15.1.

3. Review with students the criteria listed in the Persuasive Speech Assessment Form (Handout 15.2).

4. Have students work on their presentation (this may take 2 or 3 sessions).

THINGS TO ASK

- What parts of the design do you think people ought to see?
- What information is critical for them to know in order to understand the positive and negative aspects of the design?
- Why is this information important?
- Which aspects (positive and negative) do we want to highlight? Do we need to highlight any negative aspects?
- Which aspects of design would be best presented visually? Which through oral presentations?

Session 4

THINGS TO DO

1. If possible, a panel of interested professionals representing various parties involved should be invited to class to hear students' design presentations. The panel may include community members who were consulted during the unit, or community members whose job might involve project review and consideration. If getting community members is not possible, perhaps assembling parents or school personnel who will assume the role of a community member (introductions would include a title and position in the community) would be an option.

2. Set up room for presentation to the City Council. Have teams present their design. Each team should have an equal amount of time to present and an opportunity to answer questions from the panel and the rest of the classroom students following the presentation.

3. The teacher needs to complete Handout 15.2 on each student.

Session 5

THINGS TO DO

1. After the guest panel has left, have students as a class discuss the designs presented. Discuss both positive and negative aspects.
2. Have students complete Handout 15.3.

ASSESSMENT

1. Group presentations.
2. Evaluation of Persuasive Speech (Handout 15.2).
3. Problem Log Questions (Handout 15.3).

NOTE TO TEACHER

Coach your panelists to ask questions of the students so that students must defend their design.

HANDOUT 15.1

PRESENTATION PLANNING SHEET

Name: _____

Presentation Topic: _____

Things I know about this topic:

The "big ideas" to get across to the audience:

The connections I want to make:

What are some interesting ways I can present this information to others?

What materials will I need in order to best present my information?

HANDOUT 15.2

PERSUASIVE PRESENTATION ASSESSMENT FORM

Name: _____ Date: _____

Use the following rating scale to evaluate each quality:
  1 = Needs Improvement
  2 = Adequate
  3 = Excellent

_____ The purpose of the presentation was clear.

_____ The speaker's reasoning was clear and logical.

_____ The basic components of the proposal were evident.

_____ The speaker showed knowledge of the subject.

_____ The speaker held the interest of the audience.

_____ The speaker was audible, maintained eye contact, and spoke with expression.

The best part of this presentation was:

A suggestion for improvement is:

HANDOUT 15.3
PROBLEM LOG QUESTIONS

1. What are the criteria that you will use to judge the success of your design? Briefly describe why each criterion represents an important concern in the development of the design.

   a.

   b.

   c.

   d.

   e.

   f.

   g.

   h.

   i.

   j.

2. Look at your criteria and look at the design your group came up with. Did all of the criteria receive equal weight in the design? Did some criteria seem to work against each other? What sacrifices did you have to make in order to come up with the "best possible design"? Was your "best possible" design different from your idea of an ideal design? How?

lesson

16

Final Overall
Unit Assessment Activity

INSTRUCTIONAL PURPOSE
- To assess understanding of the scientific content taught by this unit.
- To assess the ability of the student to use appropriate scientific process skills in the resolution of a real-world problem.
- To assess student understanding of the concept of systems.

ESTIMATED TIME
The content assessment should take the students approximately thirty minutes; the experimental design assessment should take the students approximately thirty minutes; and the systems assessment should take the students approximately thirty minutes.

MATERIALS AND HANDOUTS
Handout 16.1: Final Content Assessment

Handout 16.2: Experimental Design Assessment

Handout 16.3: Systems Assessment

Scoring protocols for Final Content Assessment, Experimental Design Assessment, and Systems Assessment

PROCEDURE

Have students complete assessments found in Handouts 16.1, 16.2, and 16.3.

1. Suppose you had the experimental setup diagrammed below, in which one end of a wire is connected to the positive terminal of a battery and the other end is connected to the base of a small light bulb; a second wire runs from the base of the light bulb to the battery's negative terminal; and the light bulb is shining.

   a. Is the circuit in the experimental setup a complete circuit or an incomplete circuit? How do you know?

   Suppose you made the following change in your experimental setup: you cut wire number two, stripped the insulation off the ends with a wire stripper, and then placed the two ends in a cup full of salt water in such a way that the two ends did not touch, as shown in the diagram below. After you did this, you noticed that your light bulb still shone.

   b. Based on this experiment, is salt water a conductor or an insulator? Explain your answer.

   Bowl of Salt Water

   c. Saliva is salty. Based on this and on your answers to parts a and b, why is it a bad idea to lick an electrical outlet?

2. Suppose you had three bulbs, a battery, and lots of wires.

   a. Using the bulbs, wires, and battery, how could you make a complete circuit in which the resistors were in series? Using appropriate electrical symbols, draw a diagram of this circuit.

   b. Using the bulbs, wires, and battery, how could you make a complete circuit in which the resistors were in parallel? Using appropriate electrical symbols, draw a diagram of this circuit.

3. Is a light bulb an example of a resistor, a battery, or an insulator? Explain your answer.

4. The following diagram shows a complete circuit.

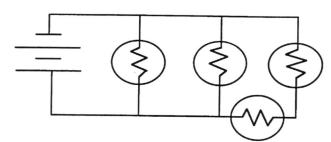

   a. In red, draw an arrow that shows the direction in which current flows in this circuit.
   b. In blue, draw an arrow that shows the direction in which electrons flow in this circuit.

5. Suppose you were out riding your bike after a storm and came upon a power line that had been knocked to the ground by a tree.

   a. Who should you call for help?

   b. What should you do (or have your parents or other adults do) to make sure that nobody gets hurt before help arrives?

HANDOUT 16.2

EXPERIMENTAL DESIGN ASSESSMENT *(30 MINUTES)*

Your uncle, who is a bit of a tease, tells you that potatoes can conduct electricity. You have potatoes, batteries, wire, and some small light bulbs with which you can make circuits.

What experiment could you do that would allow you to see if what your uncle says is true? In your answer, include the following:

a. Your hypothesis:

b. The materials you would need, including any necessary safety equipment:

c. The protocol you would use:

d. A data table showing what data you would collect:

e. A description of how you would use your data to decide whether potatoes are conductors:

HANDOUT 16.3

SYSTEMS ASSESSMENT (30 MINUTES)

The electrical wiring in your home is a system. For this system, do the following:

1. List the parts of the system in the spaces provided below. Include boundaries, elements, input, and output.

    Boundaries (describe):

    Elements (list at least five):

    Input (list at least two kinds):

    Output (list at least two kinds):

2. Draw a diagram of the system in question #1, showing where each of the parts can be found.

3. On your diagram, draw lines (in a different color) showing three important interactions between different parts of the system. Why is each of these interactions important to the system? Explain your answer.

   a. Interaction #1:

   b. Interaction #2:

   c. Interaction #3:

SCORING PROTOCOL
FINAL CONTENT ASSESSMENT

1. **(15 point total)** Suppose you had the experimental setup diagrammed below, in which one end of a wire is connected to the positive terminal of a battery and the other end is connected to the base of a small light bulb; a second wire runs from the base of the light bulb to the battery's negative terminal; and the light bulb is shining.

a. **(5 points)** Is the circuit in the experimental setup a complete circuit or an incomplete circuit? How do you know?

The circuit is complete, because the light bulb shines thanks to the charge in the battery; if it were incomplete, the light bulb would stay off.

Scoring: Give five points for the answer and explanation.

Suppose you made the following change in your experimental setup: you cut wire number two, stripped the insulation off the ends with a wire stripper, and then placed the two ends in a cup full of salt water in such a way that the two ends did not touch, as shown in the diagram below. After you did this, you noticed that your light bulb still shone.

Bowl of Salt Water

b. **(5 points)** Based on this experiment, is salt water a conductor or an insulator? Explain your answer.

Salt water is a conductor, because it can form a part of a complete circuit. If it were an insulator, no charge would flow and the bulb would not shine.

Scoring: Give five points for the answer and explanation.

c. **(5 points)** Saliva is salty. Based on this and on your answers to parts a and b, why is it a bad idea to lick an electrical outlet?

The saliva on your tongue could run into the outlet, making your body an inadvertent part of a complete circuit that had house current running through it; this could be deadly.

Scoring: Give five points for the answer and explanation.

2. **(10 point total)** Suppose you had three bulbs, a battery, and lots of wires.

   a. **(5 points)** Using the bulbs, wires, and battery, how could you make a complete circuit in which the resistors were in series? Using appropriate electrical symbols, draw a diagram of this circuit.

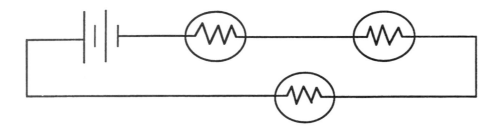

   b. **(5 points)** Using the bulbs, wires, and battery, how could you make a complete circuit in which the resistors were in parallel? Using appropriate electrical symbols, draw a diagram of this circuit.

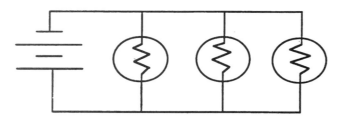

3. **(5 points)** Is a light bulb an example of a resistor, a battery, or an insulator? Explain your answer.

   A light bulb is an example of a resistor, as its presence in the circuit impedes current flow; it lights through a phenomenon known as resistance heating, in which current running through a conductor causes it to heat up, and, in the case of the filament of a light bulb, emit light.

Scoring: Give five points for any reasonable answer and explanation.

4. **(10 point total)** The following diagram shows a complete circuit.

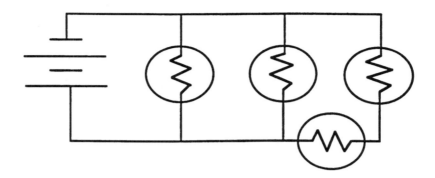

a. **(5 points)** In red, draw an arrow that shows the direction in which current flows in this circuit.

b. **(5 points)** In blue, draw an arrow that shows the direction in which electrons flow in this circuit.

Current flows from the positive terminal to the negative terminal of the battery (thanks to Ben Franklin's arbitrary definition of current, made before the electron had been discovered!); electrons flow in the opposite direction.

5. **(10 point total)** Suppose you were out riding your bike after a storm and came upon a power line that had been knocked to the ground by a tree.

a. **(5 points)** Who should you call for help?

You or your parents should call the police, to tell them that there is a line down so that they can protect people from it; you or the police should call the power company, to let them know that a line is down. Accept any reasonable answer.

b. **(5 points)** What should you do (or have your parents or other adults do) to make sure that nobody gets hurt before help arrives?

An adult should make sure that nobody touches the power line before help arrives.

Accept any reasonable answer.

Total number of points possible: 50

SCORING PROTOCOL
EXPERIMENTAL DESIGN ASSESSMENT

Your uncle, who is a bit of a tease, tells you that potatoes can conduct electricity. You have potatoes, batteries, wire, and some small light bulbs with which you can make circuits. What experiment can you do to see if what your uncle says is true?

In your answer, include the following:

a. **(5 points)** Your hypothesis:

Potatoes are conductors. **Note:** *Other hypotheses are possible; accept all reasonable answers; give five points for any reasonable hypothesis.*

b. **(10 points)** The materials you would need, including any necessary safety equipment:

Batteries
Bulbs
Potatoes
Wire

Note: The student only needs to list the above; this materials list does not need to be comprehensive. Accept all reasonable materials lists, as long as they're consonant with the hypothesis in part a.

c. **(10 points)** The protocol you would use:

I would first do a control experiment to make sure that all of my circuit elements worked. I would make a complete circuit containing the batteries and bulb that I planned to use for the potato experiment. If the bulb lit up, I would know that all of my circuit elements worked as expected.

The second part of my experiment would involve testing the conductive properties of the potato. I would use the control circuit set up in the first experiment and add the potato to the circuit by cutting one of the control circuit's wires, stripping the ends, and inserting each end into the potato.

The final part of my experiment would involve removing the potato, cleaning the cut ends of the wire that had been inserted into the potato, and attaching them to each other in such a way as to assure good electrical contact. I would do this to be sure that the circuit elements hadn't failed during the potato experiment. If the bulb still lit, I would know that any failure in the potato experiment had to be due to something about the potato.

Scoring: Give five points for any reasonable protocol (or experimental outline: not every step needs to be listed in fine detail, but it should be clear what the student intends to test) that it is consonant with the hypothesis given in part a (if the two seem to be unrelated, withhold these points); give five points for the presence of a control for the experiment.

    d. **(15 points)** A data table showing what data you would collect

Experiment	_Bulb lit?_
Pre-control	
Potato experiment	
Post-control	

Scoring: Give five points for the presence of a data table; 5 points if there is an independent variable (not necessarily labeled as such) present in the data table headings; and 5 points if there is at least one dependent variable (not necessarily labeled as such) present in the data table headings. In this answer, the presence or absence of the potato is the independent variable, and whether the bulb lit is the dependent variable.

Note: Accept all reasonable answers, as long as they're consonant with the student's answers to parts a–d.

    e. **(10 points)** A description of how you would use your data to decide whether potatoes are conductors.

> _If the light lit in both my pre and post control experiments, I would know that the behavior of the circuit in the potato experiment was solely dependent on the properties of the potato. If the bulb lit in this experiment, I would know that the potato was a conductor; if it failed to light, I would know that the potato was not a conductor._

Scoring: Give ten points for an answer that explains how the data will be used to come up with a conclusion. If the student doesn't mention the data, then give no points.

Note: Accept all reasonable answers, as long as they're consonant with the student's answers to parts a–d.

Total number of points possible: 50

SCORING PROTOCOL
SYSTEMS ASSESSMENT

The electrical wiring in your home is a system. For this system, do the following:

1. **(25 points)** List the parts of the system. Include boundaries, elements, input, and output.

    Boundaries (describe):

    The outside boundaries of the home, not including the outside line that goes to the power line in the street.

    Scoring: For ten points, accept any reasonable <u>closed</u> boundaries.

    Elements (list at least five):

    Wires, fuses, fuse box, outlets, outlet covers, electrical appliances, etc.

    Scoring: Give one point for each reasonable element consistent with the boundaries up to a maximum of five points.

    Input (list at least two kinds):

    Electricity from the power grid; corrosion of the wires that comes from age; gnawing by squirrel teeth in the attic; new appliances that are plugged in and used.

    Scoring: Give 2.5 points for each listed input item consistent with the boundaries up to a maximum of five points.

    Output (list at least two kinds):

    Light from light bulbs; heat from electric water heaters, toasters, stoves; noise from the TV and radio.

    Scoring: Give 2.5 points for each listed output item consistent with the boundaries, up to five points.

2. **(10 points)** Draw a diagram of the system that shows where each of the parts can be found.

    Scoring: Accept any reasonable diagram that includes the system components listed in the answer to question 1.

3. **(15 points)** On your diagram, draw lines (in a different color) showing three important interactions between different parts of the system. Why is each of these interactions important to the system? Explain your answer.

    a. Interaction #1:

    Squirrel saliva and wires could create a new circuit, resulting in a loss of power and a fried squirrel.

b. Interaction #2:

Electricity running through the filament of a light bulb results in light from the filament.

c. Interaction #3:

Electricity running through the TV when it is turned on results in lights and noises.

(Accept any reasonable interaction; give five points for each correct answer.)

Total number of points possible: 50

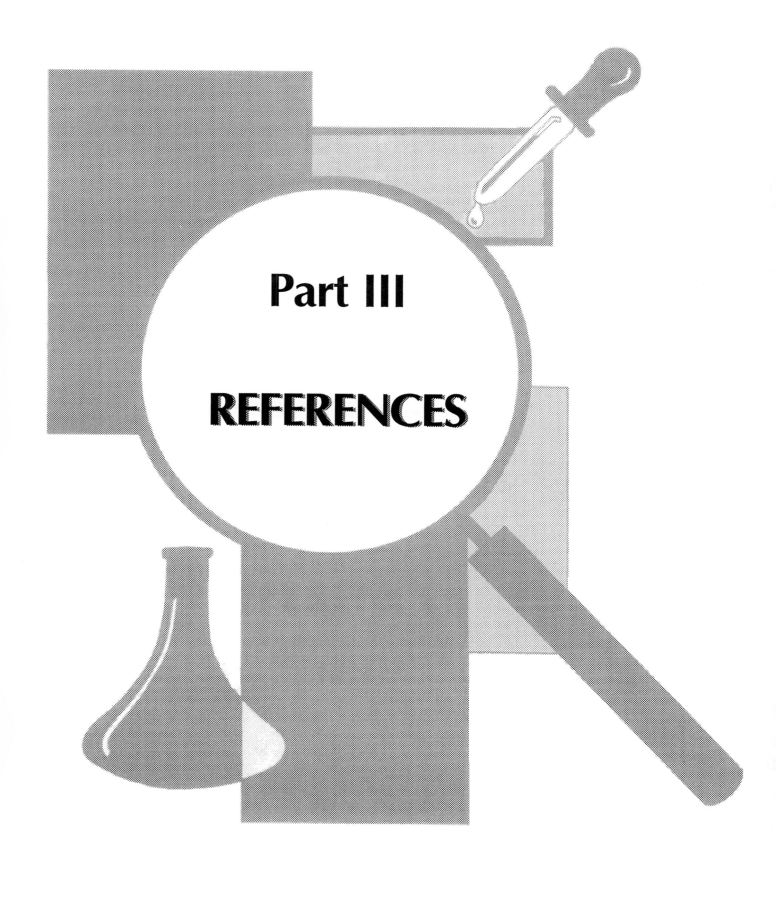

Part III

REFERENCES

REFERENCES

Ardley, N. (1991). *The science book of electricity.* NY: Harcourt Brace Jovanovich.

Beasant, P., & Smith, A. (1993). *How to draw maps and charts.* Tulsa, OK: EDC Publishing (Usborne).

Cothron, J.H., Giese, R.N., & Rezba, R.J. (1996). *Science experiments and projects for students.* Dubuque, IA: Kendall/Hunt Publishing Company.

Cothron, J.H., Giese, R.N., & Rezba, R.J. (1996). *Science experiments by the hundreds.* Dubuque, IA: Kendall/Hunt Publishing Company.

Cothron, J.H., Giese, R.N., & Rezba, R.J. (1996). *Students and research: Practical strategies for science classrooms and competition.* Dubuque, IA: Kendall/Hunt Publishing Company.

Farrow, S. (1996). *The really useful science book: A framework of knowledge for primary teachers.* London: Falmer Press.

Livingston, J.D. (1996). *Driving force: The natural magic of magnets.* Cambridge, MA: Harvard University Press.

Markel, S. (1989). *Power up: Experiments, puzzles, and games exploring electricity.* NY: Macmillan Publishing Company.

National Science Resources Center (1991). *Science and Technology for Children.* Electric Circuits Unit. Washington, DC: National Academy of Sciences.

Parker, S. (1992). *Electricity.* NY: Dorling Kindersley, Inc.

Parker, S. (1992). *Science discoveries: Thomas Edison and electricity.* NY: Chelsea House Publishers.

Snedden, R. (1995). *The history of electricity.* NY: Thomas Learning.

Vogt, G. (1986). *Generating electricity.* NY: Franklin Watts.

Ward, A. (1991). *Experimenting with batteries, bulbs, and wires.* NY: Chelsea House Publishers.

Whyman, K. (1989). *Sparks to power stations.* NY: Gloucester Press.

Order these outstanding titles by the
CENTER FOR GIFTED EDUCATION

SCIENCE

QTY	TITLE	ISBN	PRICE	TOTAL
	Guide to Teaching a Problem-Based Science Curriculum	0-7872-3328-5	$32.95*	
	Acid, Acid Everywhere	0-7872-2468-5	$32.95*	
	The Chesapeake Bay	0-7872-2518-5	$32.95*	
	Dust Bowl	0-7872-2754-4	$32.95*	
	Electricity City	0-7872-2916-4	$32.95*	
	Hot Rods	0-7872-2813-3	$32.95*	
	No Quick Fix	0-7872-2846-X	$32.95*	
	What a Find!	0-7872-2608-4	$32.95*	

LANGUAGE ARTS

QTY	TITLE	ISBN	PRICE	TOTAL
	Guide to Teaching a Language Arts Curriculum for High-Ability Learners	0-7872-5349-9	$32.95*	
	Autobiographies Teaching Unit	0-7872-5338-3	$28.95*	
	Literature Packets	0-7872-5339-1	$37.00*	
	Journeys and Destinations Teaching Unit	0-7872-5167-4	$28.95*	
	Literature Packets	0-7872-5168-2	$37.00*	
	Literary Reflections Teaching Unit	0-7872-5288-3	$28.95*	
	Literature Packets	0-7872-5289-1	$37.00*	
	The 1940s: A Decade of Change Teaching Unit	0-7872-5344-8	$28.95*	
	Literature Packets	0-7872-5345-6	$37.00*	
	Persuasion Teaching Unit	0-7872-5341-3	$28.95*	
	Literature Packets	0-7872-5342-1	$37.00*	
	Threads of Change in 19th Century American Literature Teaching Unit	0-7872-5347-2	$28.95*	
	Literature Packets	0-7872-5348-0	$37.00*	

Method of payment:

❏ Check enclosed (payable to Kendall/Hunt Publishing Co.)

❏ Charge my credit card:

  ❏ VISA   ❏ Master Card   ❏ AmEx

AL, AZ, CA, CO, FL, GA, IA, IL, IN, KS, KY, LA, MA, MD, MI, MN, NC, NJ, NM, NY, OH, PA, TN, TX, VA, WA, & WI add sales tax.

Add shipping: order total $26-50 = $5; $51-75 = $6; $76-100 = $7; $101-150 - $8; $151 or more = $9

Price is subject to change without notice.   **TOTAL**

Credit Card No. _____

Exp. Date _____

Signature _____

Name _____

Address _____

City/State/ZIP _____

Phone No. (   ) _____

E-mail _____

KENDALL/HUNT PUBLISHING COMPANY
4050 Westmark Drive   P.O. Box 1840   Dubuque, Iowa 52004-1840
A16/mkk   Q2   2005   01

Call (800) 228-0810 • Fax (800) 772-9165
Visit us at www.kendallhunt.com

An overview of the outstanding titles available from the

CENTER FOR GIFTED EDUCATION

SCIENCE

A PROBLEM-BASED LEARNING SYSTEM FROM THE CENTER FOR GIFTED EDUCATION FOR YOUR K-8 SCIENCE LEARNERS

The Center for Gifted Education has seven curriculum units containing different real-world situations that face today's society, plus a guide to using the curriculum. The units are geared towards different elementary levels, yet can be adapted for use in all levels of K-8.

The goal of each unit is to allow students to analyze several real-world problems, understand the concept of systems, and conduct scientific experiments. These units also allow students to explore various scientific topics and identify meaningful problems for investigation.

Through these units your students experience the work of real science in applying data-handling skills, analyzing information, evaluating results, and learning to communicate their understanding to others.

LANGUAGE ARTS

A LANGUAGE ARTS CURRICULUM FROM THE CENTER FOR GIFTED EDUCATION FOR YOUR GRADES 2-11

The Center for Gifted Education at the College of William and Mary has developed a series of language arts curriculum units for high-ability learners.

The goals of each unit are to develop students' skills in literature interpretation and analysis, persuasive writing, linguistic competency, and oral communication, as well as to strengthen students' reasoning skills and understanding of the concept of change.

The units engage students in exploring carefully selected, challenging works of literature from various times, cultures, and genres, and encourage students to reflect on the readings through writing and discussion.

The units also provide numerous opportunities for students to explore interdisciplinary connections to language arts and to conduct research around issues relevant to their own lives. A guide to using the curriculum is also available.